THE FIRST PLANT-EATING DINOSAURS

Plant-eating dinosaurs (herbivores) were the real giants of the Age of the Dinosaurs. They numbered the mighty *Diplodocus* and *Seisomosaurus* among their ranks, the largest animals ever to walk the Earth. Herbivorous reptiles are known from the early part of the Age of Reptiles 350 million years ago. The first herbivorous dinosaurs evolved in the late Triassic period, appearing about the same time as the carnivores. Because the hipbones in both groups of dinosaurs are similarly formed, we know that the plant-eaters must have evolved from meat-eating dinosaurs.

PANGAEA

The world was very different in late Triassic/early Jurassic times. All the continental landmasses were joined together in one area, called Pangaea. This meant that animals of the same kind could migrate everywhere. This is why we find the remains of almost identical animals all over the world, from Australia to America.

PLATEOSAURUS

The first plant-eating dinosaurs belonged to the prosauropod group. *Plateosaurus* was a typical prosauropod. It had a long neck and small head, but perhaps its most important feature was its big body. To process plant matter a herbivore needs a far greater volume of digestive gut than a carnivore. The prosauropod's heavy mass of intestines, carried well forward of the hips, would have made the animal too unbalanced to spend much time on its hind legs, so prosauropods became four-footers early in their history.

TRIASSIC 245-208 MYA	EARLY/MID JURASSIC 208-157 MYA	LATE JURASSIC 157-146 MYA	EARLY CRETACEOUS 146-97 MYA	LATE CRETACEOUS 97-65 MYA

HELPLESS PREY

Palaeontologists have found the remains of a prosauropod *Euskelosaurus* in late Triassic rocks in South Africa and Switzerland. The bones of its feet and legs are preserved, but the rest of the skeleton is broken up and scattered. Teeth of crocodile-like reptiles and carnivorous dinosaurs are among them. From this we suppose that *Euskelosaurus* became stuck in mud, and while struggling helplessly, was attacked by meat-eaters.

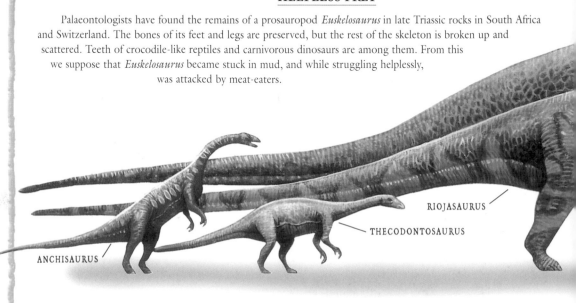

RIOJASAURUS

THECODONTOSAURUS

ANCHISAURUS

LIFE OF THE PROSAUROPODS

Ever since plants have existed plant-eating animals have fed on them, and ever since plant-eating animals have existed meat-eaters have, in turn, fed on them. This type of food chain can still be seen today on the grasslands of Africa where herds of herbivorous wildebeest and zebra graze on the low vegetation of the savannah and, in turn, are preyed upon by prowling carnivores such as lions and cheetahs. It was no different during the age of the dinosaurs. Prosauropods fed on the trees and were themselves stalked by the meat-eaters of the time, including the early carnivorous dinosaurs.

TRACKWAYS

The footprints called *Navajopus* from early Jurassic rocks of Arizona perfectly match the foot bones of a typical prosauropod, with the big hind feet and smaller front feet, each with four toes and inwardly-curved claws. They are likely to have been made by a small *Thecodontosaurus*-sized prosauropod called *Ammosaurus*.

MELANOROSAURUS

RANGE OF PROSAUROPODS

During the Triassic period, Sauropods ranged all over Pangaea, the world's landmass. *Melanorosaurus* lived in South Africa, *Thecodontosaurus* in western Europe, *Anchisaurus* in western North America and *Riojasaurus* in South America. Other prosauropods, such as *Plateosaurus* and the Plateosaurus-like *Lufengosaurus* lived in what is now China. They were all extinct by middle Jurassic times.

TRIASSIC 245-208 MYA	EARLY/MID JURASSIC 208-157 MYA	LATE JURASSIC 157-146 MYA	EARLY CRETACEOUS 146-97 MYA	LATE CRETACEOUS 97-65 MYA

SAUROPODS

STOMACH STONES

The small head and mouth of sauropods were not designed for chewing. To help break down the food sauropods swallowed stones, which ground up plant material as it passed. We know this because gastroliths (stomach stones) have been found among their bones. Today, many plant-eating birds do the same.

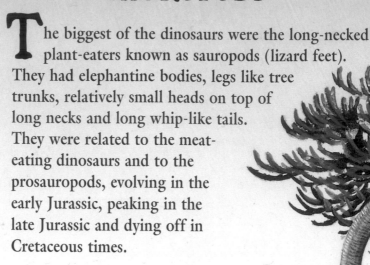

The biggest of the dinosaurs were the long-necked plant-eaters known as sauropods (lizard feet). They had elephantine bodies, legs like tree trunks, relatively small heads on top of long necks and long whip-like tails. They were related to the meat-eating dinosaurs and to the prosauropods, evolving in the early Jurassic, peaking in the late Jurassic and dying off in Cretaceous times.

DIPLODOCUS

Perhaps the best known of the long sauropods is *Diplodocus*. At 27 metres (88 ft) long, it was one of several sauropods that roamed the plains and woodlands of North America in late Jurassic times. The way the neck bones were articulated tells us they browsed on low ferny vegetation, probably sweeping out great arcs with their long necks. The balance of muscles at the hips would also have enabled them to stretch up on their hind legs to reach the trees. The signs of wear on their teeth show they fed in both positions.

SAUROPOD FRAME

Remains of sauropod skeletons consist of massive pieces of fossilized bone, so big that there is nothing alive

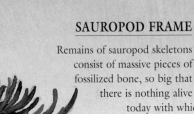

today with which scientists can compare them. In one of the latest techniques, very basic bone shapes are programmed into a computer and manipulated to let us see how the various pieces moved against one another. These studies are particularly useful for studying the flexibility of necks and tails.

TRACES OF LIFESTYLE

We used to think sauropods were too heavy to spend much time on land, and must have supported their vast bulk by wading in deep water. However, we now know (mostly from fossilized footprints) that sauropods moved about in herds on dry land. Large and small footprints found together show that different sauropods lived in groups. As there is no sign of tail marks in the tracks, they must have kept their tails raised.

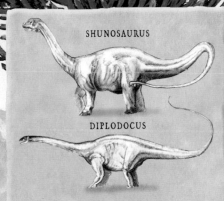

SHUNOSAURUS

DIPLODOCUS

IN DEFENCE

Sauropods would have been prey to the big carnivorous dinosaurs. Just as today tigers do not attack fully-grown elephants, in Jurassic times the biggest of the sauropods would have been safe from the meat-eaters but the young and the sick would have been under constant threat. *Diplodocus* probably protected itself and its herd by using its long tapering tail as a whip. *Shunosaurus*, which lived in China during the middle Jurassic, probably used the small club on the end of its tail to defend itself.

| TRIASSIC 245-208 MYA | EARLY/MID JURASSIC 208-157 MYA | LATE JURASSIC 157-146 MYA | EARLY CRETACEOUS 146-97 MYA | LATE CRETACEOUS 97-65 MYA |

APATOSAURUS GROWTH RATE

It is difficult to tell how long a dinosaur lived. Sometimes, growth lines in the bones (like the rings of trees) suggest the animal grew more quickly at some time of each year. Its age can be assessed by counting the lines. Studies of the bones of *Apatosaurus* (previously known as *Brontosaurus*), the relative of *Diplodocus*, suggest that these sauropods grew quickly, without growth rings, for about 10 years. By then, they had reached 90 per cent of their adult size. After they reached this age, they grew very little.

10 YEARS

BRACHIOSAURUS

Although many remains have been found in the Morrison Formation, the best skeleton of *Brachiosaurus* was found halfway across the globe in Tanzania. This shows that in late Jurassic times Pangaea (*see pages 2–3*) had not yet split up completely and the same types of dinosaur lived all over the world. A German expedition unearthed this skeleton in 1909, when Tanzania was still called German East Africa. The complete skeleton, the biggest mounted anywhere, is in the Humboldt Museum in Berlin.

TRIASSIC 245-208 MYA	EARLY/MID JURASSIC 208-157 MYA	LATE JURASSIC 157-146 MYA	EARLY CRETACEOUS 146-97 MYA	LATE CRETACEOUS 97-65 MYA

THE HEYDAY OF THE SAUROPODS

During the late Jurassic period, sauropods were at their most prolific and widespread. Some were long and low, and browsed the low vegetation. Others were tall and browsed the lower branches of trees. There were two main types, distinguished by the shape of their teeth. *Diplodocus* and the other long, low sauropods had peg-like teeth; while the taller stouter sauropods, such as *Brachiosaurus*, had thick, spoon-shaped teeth which indicate a different type of feeding arrangement. However, nobody is yet sure what it was.

DINOSAUR DETECTIVES

Often, when the remains of a very big animal are discovered there are tantalizingly few bones found. Comparing them directly to a more complete skeleton can give us some idea of the kind of animal they came from. In 1999, four neck vertebrae of a gigantic sauropod turned up. The *Sauroposeidon* bones turned out to be very similar to the neck bones of *Brachiosaurus*. So, we are fairly sure *Sauroposeidon* was an animal very like *Brachiosaurus* – but bigger!

SEISMOSAURUS

The longest dinosaur known is *Seismosaurus*. Imagine *Diplodocus*, then double its length. Make this length by stretching the neck and the tail in proportion to the body and this is what *Seismosaurus* looked like. So far, only one *Seismosaurus* skeleton has been found and that was in the Morrison Formation rocks in New Mexico. The skeleton is of an animal that may have been about 50 metres (164 ft) long.

DRESSED TO IMPRESS

Not only did the late sauropods have armour, but some of them had strange spines and frills as well. *Amargasaurus* from early Cretaceous Argentina had a double row of spines down its neck and a tall fin down its back. Unusual sauropods evolved in Cretaceous South America because it was an island continent at the time, and evolution took an independent direction there.

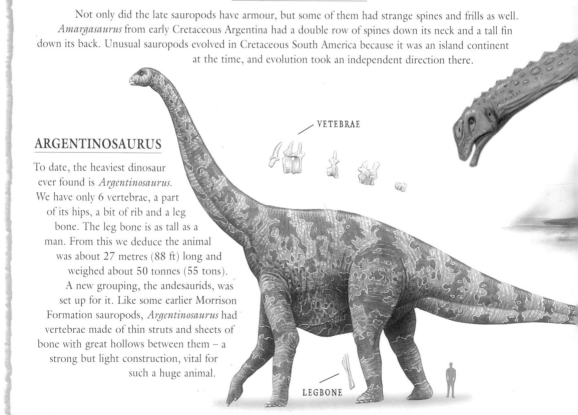

ARGENTINOSAURUS

To date, the heaviest dinosaur ever found is *Argentinosaurus*. We have only 6 vertebrae, a part of its hips, a bit of rib and a leg bone. The leg bone is as tall as a man. From this we deduce the animal was about 27 metres (88 ft) long and weighed about 50 tonnes (55 tons). A new grouping, the andesaurids, was set up for it. Like some earlier Morrison Formation sauropods, *Argentinosaurus* had vertebrae made of thin struts and sheets of bone with great hollows between them – a strong but light construction, vital for such a huge animal.

VETEBRAE

LEGBONE

THE LAST OF THE SAUROPODS

As the world passed from the Jurassic into the Cretaceous period, the vegetation began to change and the continents to move apart. Different dinosaurs were becoming prominent. The sauropods began to die away as a completely different group of plant-eating dinosaurs evolved. In some places the sauropods still thrived, either because the old style vegetation still flourished in some environments, or because they lived on isolated continents where the new dinosaurs did not reach. Despite the spread of the new dinosaur types there were sauropods existing right to the very end of the Dinosaur Age.

TOUGH GUY

The bony armour pieces from the back of a titanosaurid were found as long ago as 1890 in Madagascar. The palaeontologist who first identified them was not believed, as no other sauropod was known to be covered in armour. Only with the discovery of armoured titanosaurids in Argentina in the 1970s and a more complete armoured titanosaurid in Madagascar in the 1990s was this scientist's theory proven to be correct.

SALTASAURUS

Of the sauropods that survived into the Cretaceous period the titanosaurids (such as *Saltasaurus* in Argentina or *Ampelosaurus* in France) were perhaps the most successful. Despite their name, at about 12 metres (39 ft) long, they were not particularly big for sauropods. In recent years, it has been found that at least some titanosaurids had a back covered in armour. This may not have been for defence but, like the shell on the back of a crab, it might have been for stiffening the backbone to help the animal carry its weight.

TRIASSIC	EARLY/MID JURASSIC	LATE JURASSIC	EARLY CRETACEOUS	LATE CRETACEOUS
245-208 MYA	208-157 MYA	157-146 MYA	146-97 MYA	97-65 MYA

ORNITHOPODS - THE BIRD FEET

HYPSILOPHODON SKULL

The skull of an ornithopod was different from that of a sauropod. There was always a beak at the front for cropping the food. The teeth were not merely for raking in leaves but were designed for chewing them, either by chopping or grinding. Depressions at each side of the skull show where there were probably cheek pouches, used to hold the food while it was being processed. This is a far more complicated arrangement than that of the prosauropods and sauropods.

During the Triassic period, at about the same time as the meat-eating dinosaurs and the prosauropods appeared, another group of plant-eating dinosaurs developed. What made these dinosaurs different was the arrangement of bones in the hip, which gave more space for the big gut needed by plant-eaters, yet enabled them to balance on their hind legs. As a result, most of these animals were two-footed herbivores. Victorian scientists called the long-necked plant-eaters the sauropods (lizard feet) because they had a lizard-like arrangement of bones in their feet; the two-footed, bird-hipped dinosaurs they called the ornithopods (bird feet). We still use these terms today.

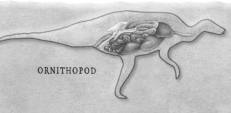

ADVANCED JAWS

Later, more advanced ornithopods had quite complex chewing mechanisms. An animal like *Iguanodon* (*see page 15*) or a hadrosaur (*see pages 16–17*) had its upper teeth mounted on articulated plates at each side of the skull. As the lower jaw rose these plates moved outwards to allow the sloping chewing surfaces of both sets of teeth to grind past one another. This constant milling action wore away the teeth, and new ones grew to replace them.

SAUROPOD

ORNITHOPOD

HIPBONES

As with the prosauropods (*see pages 4–5*) the hipbones of the sauropods incorporated a pubic bone that pointed down and forward. This meant the big plant-digesting intestines had to be carried forward of the hips. In ornithopods this pubic bone is swept back, except for a pair of forward extensions that splayed out to the side. The big plant-digesting intestines could be carried beneath the animal's hipbone, closer to its centre of gravity. This enabled the ornithopod to walk on its hind legs, balanced by its tail – just like a meat-eating dinosaur.

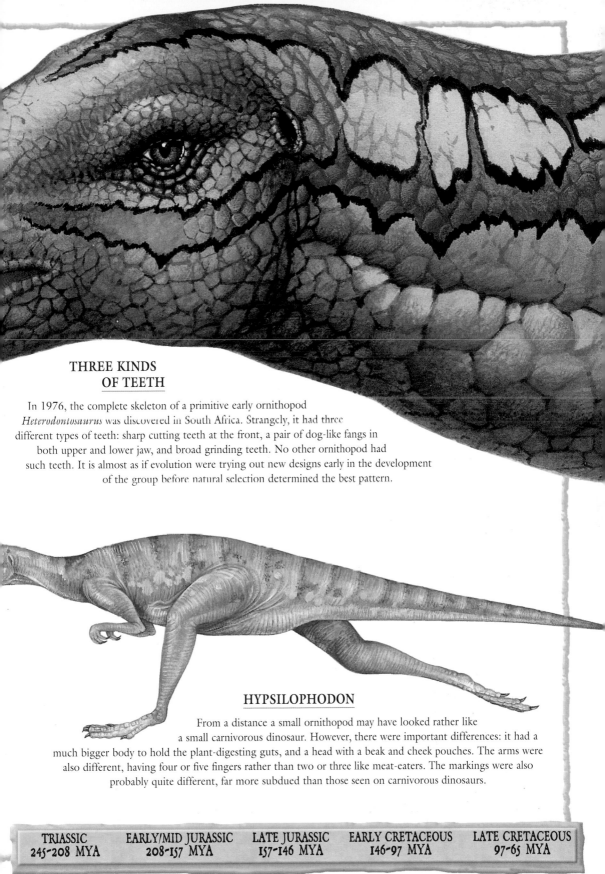

THREE KINDS
OF TEETH

In 1976, the complete skeleton of a primitive early ornithopod
Heterodontosaurus was discovered in South Africa. Strangely, it had three
different types of teeth: sharp cutting teeth at the front, a pair of dog-like fangs in
both upper and lower jaw, and broad grinding teeth. No other ornithopod had
such teeth. It is almost as if evolution were trying out new designs early in the development
of the group before natural selection determined the best pattern.

HYPSILOPHODON

From a distance a small ornithopod may have looked rather like
a small carnivorous dinosaur. However, there were important differences: it had a
much bigger body to hold the plant-digesting guts, and a head with a beak and cheek pouches. The arms were
also different, having four or five fingers rather than two or three like meat-eaters. The markings were also
probably quite different, far more subdued than those seen on carnivorous dinosaurs.

TRIASSIC 245-208 MYA	EARLY/MID JURASSIC 208-157 MYA	LATE JURASSIC 157-146 MYA	EARLY CRETACEOUS 146-97 MYA	LATE CRETACEOUS 97-65 MYA

CHANGING FACE

Over the years, as more specimens were found, *Iguanodon's* appearance changed. In the 1850s, it was constructed in the Crystal Palace gardens, south London, along the lines of Mantell's big lizard. Then, in 1878, a whole herd of *Iguanodon* skeletons, mostly complete, were found in a coal mine in Bernissart, Belgium. These animals were up to 10 metres (33 ft) long and had hind legs that were much longer than their forelimbs.

This evidence led to reconstructions of *Iguanodon* sitting on its hind legs, resting on its tail like a kangaroo – an image that was accepted for the next century.

IGUANA TOOTH

The first remains of *Iguanodon* – teeth and parts of bones – were discovered in Kent in about 1822 by English country doctor Gideon Mantell and his wife Mary. Other scientists of the day thought they were the teeth of fish, or of a hippopotamus. But Mantell realized the teeth were from a plant-eating reptile like a modern iguana lizard. His first reconstructions showed a kind of a dragon-sized, iguana-like reptile, similar to the first reconstructions of the meat-eating *Megalosaurus*, also recently discovered.

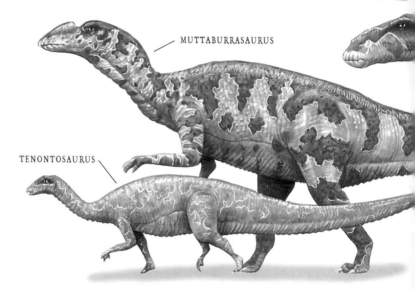

MUTTABURRASAURUS

TENONTOSAURUS

THE OURANOSAURUS QUESTION

One iguanodontid, *Ouranosaurus*, had an arrangement of tall spines forming a kind of a wicket fence along its backbone, probably to support some sort of fin or sail. As *Ouranosaurus* lived in North Africa, which was hot and arid during Cretaceous times, such a sail could have regulated its body temperature by exposing blood vessels to the warming sun and cooling wind. A meat-eater, *Spinosaurus*, lived in the same time and place and also had a sail. Another theory is that the spines supported a fatty hump, such as camels have today.

THE IGUANODON DYNASTY

Iguanodon was among the first dinosaurs to be discovered. Its remains came to light about the same time as those of the meat-eater *Megalosaurus*. The teeth and a few scraps of bone were found first and were obviously from a very large plant-eating reptile. At the time, few people were familiar with modern plant-eating reptiles, so the animal was particularly unusual. Some scientific work was being done on the modern South American plant-eating lizard, the iguana, which had teeth rather like those of this new fossil. Hence, it was given the name, *Iguanodon*.

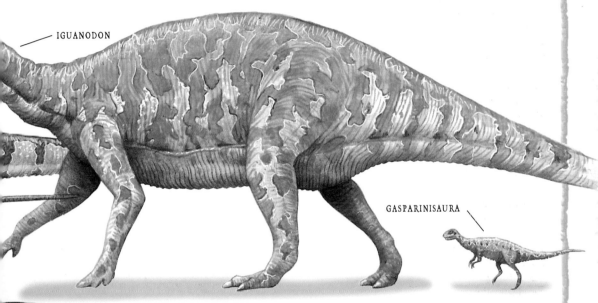

IGUANODON

GASPARINISAURA

MEET THE FAMILY

The modern interpretation of *Iguanodon* is that, as an adult, it was too heavy to spend much time on its hind legs, so went about mostly on all fours. Since it was discovered, scientists have found many more closely-related iguanodontids. Australian *Muttaburrasaurus* was slightly smaller and had a high ridge on its nose. American *Tenontosaurus* had a particularly long tail. The most primitive member of the group was *Gasparinisaura* from Argentina, which was only the size of a turkey. All of them had lived during the Cretaceous period.

IGUANODON FOOD

Iguanodon lived in northern Europe during early Cretaceous times. It wandered in herds across swampy landscapes, knee-deep in reed-beds of horsetail plants that grew just like our modern species. The herds probably grazed on these horsetails as they moved from one area to another.

TRIASSIC 245-208 MYA	EARLY/MID JURASSIC 208-157 MYA	LATE JURASSIC 157-146 MYA	EARLY CRETACEOUS 146-97 MYA	LATE CRETACEOUS 97-65 MYA

THE DUCKBILLS

In the later Cretaceous period, a new group of ornithopods evolved from the iguanodontids. The vegetation was changing: primitive forests were giving way to modern-looking woodlands of oak, beech and other broad-leaved trees with clusters of conifers and undergrowths of flowering herbs. These new dinosaurs, the hadrosaurs, spread and flourished in the broad-leaved forests throughout Europe, Asia and North America. They had thousands of grinding teeth and the front of the mouth supported a broad beak. These dinosaurs took over from the sauropods in terms of importance.

HADROSAURS

SAUROPODS

SPREAD OF DUCKBILLS

The hadrosaurs spread from Europe and became the most important plant-eating dinosaurs in the northern hemisphere. At the time, Europe, Asia and North America were joined in a single landmass and animals could spread freely. However, South America was an island continent separated from North America by a wide seaway. Hadrosaurs did reach South America but never gained a secure foothold. Long-necked sauropods remained the most important plant-eaters in South America until the end of the Dinosaur Age.

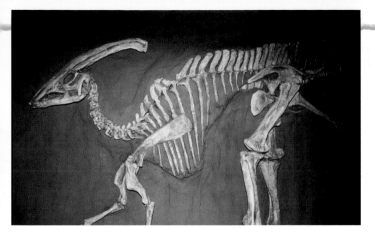

HEAD CRESTS

Some hadrosaurs, such as this *Parasaurolophus*, were quite striking because of their elaborate head crests. Mostly made of hollow bone connected to the nostrils, they were probably used for making noises to signal to one another through dense forests. Each type of hadrosaur had a unique crest shape, so that different herds could be distinguished from one another. Those with flat heads or solid crests probably supported an inflatable flap of skin that could have been puffed up like a frog's throat to make a noise.

MODERN CONIFERS

Modern conifers, such as pine and spruce, as well as the broad-leaved trees, such as oak and ash, appeared in Cretaceous times. Until then, the more primitive conifers, such as monkey puzzle trees, had sustained the sauropods. Hadrosaurs were well-equipped for dealing with the new conifer trees. They used their broad beaks to scrape off the needles and their batteries of teeth to grind them down before swallowing.

HADROSAURUS

Hadrosaurus was, like *Iguanodon*, essentially a two-footed plant-eating dinosaur which as an adult would have been rather too heavy to spend much time on its hind legs. It would have moved about on all fours, a theory confirmed by the fleshy weight-bearing pads it had on its forelimbs. *Hadrosaurus'* tail was very deep and flat which once led scientists to think the hadrosaur may have been a swimming animal – an idea that has now been discarded. Its most distinctive feature, was its broad, flat, duck-like beak.

| TRIASSIC 245-208 MYA | EARLY/MID JURASSIC 208-157 MYA | LATE JURASSIC 157-146 MYA | EARLY CRETACEOUS 146-97 MYA | LATE CRETACEOUS 97-65 MYA |

DEATH OF DENVER STEGOSAURUS

A team from the Denver Museum discovered a stegosaurus skeleton
by accident while excavating another dinosaur. This *Stegosaurus* had a
diseased tail after a broken tail spike became infected. The weakened animal
then died during a drought. Its internal organs rotted and its stomach
bloated, rolling it over on to its back. The drought ended and a nearby river
burst its banks, covering the *Stegosaurus* with silt. All this was deduced
140 million years later from the fossil and the types of rocks found nearby.
Such study of the lead-up to fossilization is known as *taphonomy*.

STEGOSAURUS

Stegosaurus lived in North America at the end of
the Jurassic. We know it was a big four-footed animal,
up to 8 metres (26 ft) long, with shorter legs at the front.
A double row of plates stuck up along its back, and two pairs
of spikes stuck out on either side towards the tip of its tail.
It had a very small head and a kind of armoured mesh
protected its throat. Some scientists
think the plates were covered in horn and
formed an armoured shield. Others insist
the plates were covered in skin and acted
as heat exchangers. On cool days, they
absorbed warmth, while on hot days, the
animal could have cooled its blood by
turning the plates into
the wind.

THE BRAINS OF
THE FAMILY?

The head of a *Stegosaurus*
was particularly small
and held a very
small brain. As with
ornithopods and
their relatives it had
a beak at the front of the
mouth and probably cheeks along the side.

| TRIASSIC 245-208 MYA | EARLY/MID JURASSIC 208-157 MYA | LATE JURASSIC 157-146 MYA | EARLY CRETACEOUS 146-97 MYA | LATE CRETACEOUS 97-65 MYA |

THE PLATED LIZARDS

Not long after the ornithopods came into existence, all kinds of other dinosaurs began to evolve from them. Many of these dinosaurs sported armour of one kind or another, and they were too heavy to spend much time on two legs. They became mostly four-footed beasts. One of the groups had armour arranged in a double row of plates or spikes down its back and tail. These plated dinosaurs were known as stegosaurs.

FLAT   PAIRS

SINGLE OVERLAPPING ROW   DOUBLE ALTERNATING

UNDER ATTACK

PLATE PUZZLES

The back plates of *Stegosaurus* were embedded in its skin but not attached directly to its skeleton. This has caused uncertainty about how they were arranged. One theory suggests the plates lay flat as armour along the animal's back. Another is that they stood upright in pairs. Yet another says they had a single upright row of overlapping plates. The most widely accepted view is that they stood in a double row, alternating with one another. Some scientists suggest the muscles at the base of the plates would have allowed *Stegosaurus* to point them at an attacker.

A WORLD OF STEGOSAURIDS

Stegosaurus was not the only stegosaurid.
There were many others, ranging from North
America, through Europe to Asia. They probably
evolved from an early Jurassic group called the
scelidosurids. The most primitive of the stegosaurids
we know were found in middle Jurassic rocks in China.
From such medium-sized animals developed the wide
range of plated and spiked dinosaurs that were important
in late Jurassic times. By the middle Cretaceous they had
all but died out. The remains of an animal that may have
been a stegosaurid was found in late Cretaceous rocks in
India. Perhaps the group lasted longer in India
which was an island continent at the time.
Or perhaps the specimen
was wrongly identified.
We are still
not sure.

KENTROSAURUS

EAST AFRICAN DISCOVERIES

The Humboldt Museum in Berlin has a collection of late Jurassic dinosaurs excavated from East Africa in the 1920s. Among them are the stegosaurid *Kentrosaurus*, which was very similar to the North American *Stegosaurus*. There were also sauropods such as *Dicraeosaurus* (shown left), which was similar to *Diplodocus*.

AN EARLY STEGOSAURID

Cow-sized *Scelidosaurus*, known from the early Jurassic rocks of England, was a four-footed herbivore covered in small studs of armour. It may have been an ancestor of the stegosaurids; or of the later Cretaceous nodosaurids of America, and the middle Jurassic to late Cretaceous ankylosaurids of Europe, North America and Asia. It may even have been ancestral to both.

VARIETY OF STEGOSAURIDS

The most primitive stegosaurid known is 4-metre (13-ft) long *Huayangosaurus* from middle Jurassic China. Later stegosaurids had shorter front legs than hind, but those of *Huayangosaurus* hardly varied. Its armour included paired narrow back spikes and a tail with two pairs of spikes. It also had a pair of shoulder spikes, as did some later stegosaurids. *Dacenturus* from late Jurassic Europe had low rounded plates on its shoulders and back, and tall spikes all down its tail. Late Jurassic *Kentrosaurus* from Africa had its centre of gravity at its hips, so, like some sauropods, it could rise on its hind legs to browse (as could *Stegosaurus*). *Wuerhosaurus* from early Cretaceous China was as big as *Stegosaurus* and had long low back plates.

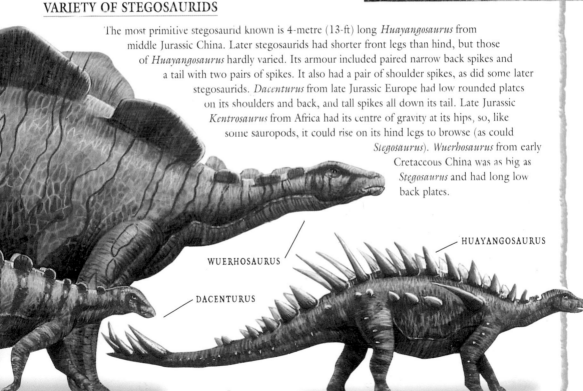

WUERHOSAURUS

DACENTURUS

HUAYANGOSAURUS

TRIASSIC 245-208 MYA	EARLY/MID JURASSIC 208-157 MYA	LATE JURASSIC 157-146 MYA	EARLY CRETACEOUS 146-97 MYA	LATE CRETACEOUS 97-65 MYA

THE NODOSAURIDS - SPIKY DINOSAURS

As the Jurassic period passed, the armoured stegosaurids became extinct and other groups of armoured dinosaurs evolved. The two most closely related groups were the nodosaurids and the ankylosaurids. Each had small bony plates across their broad backs. These plates stretched up the neck to the head and down the tail, and would have had horny covers that made the animal's back impregnable. The distinctive feature of the nodosaurid group was the presence of long robust spikes sticking out sideways and upwards from the shoulders and from the sides.

GASTONIA

One of the best-preserved nodosaurid fossils ever found was that of *Gastonia*. Its armour was tightly-packed, forming a solid shield over the hips. Spikes stood up over the shoulders, and it had a series of broad flat spines, almost blades, sticking outwards and running down each side from the neck to the tip of the tail. It was found in early Cretaceous rocks in Utah, USA, but an almost identical early Cretaceous dinosaur has been found in England.

SPIKY CUSTOMERS

Two related groups of armoured dinosaur existed in Cretaceous times. The nodosaurids were characterized by spikes on the neck and sides, while the ankylosaurids had clubs on the ends of their tails.

SAUROPELTA SKELETON

The solid back plate armour is the most commonly fossilized part of a nodosaurid, and is usually found upside-down. If a nodosaurid died and fell into a river, it may have been washed out to sea. As it decayed, expanding digestive gases in its gut would have turned it over, its heavy back acting as a keel. As it settled on the sea bed, it would be buried and eventually fossilized in that position.

STRUTHIOSAURUS

Not all nodosaurids were big animals. *Struthiosaurus* from late Cretaceous rocks of central Europe was only 2 metres (6 ½ ft) long, with a body the size of a dog. It appears to have been an island-dweller. Animals tend to evolve into smaller forms on islands to make the best use of the limited food stocks. A modern example would be the tiny Shetland pony from the islands off Scotland.

SAUROPELTA

One of the earliest nodosaurids was *Sauropelta* from Montana and Wyoming, USA. It had an arched back, long tail, and hind legs that were longer than its fore legs. Like all other nodosaurids its neck, back and tail were covered in armour. *Sauropelta's* long defensive spines were confined to the neck and shoulders, and spread outwards and upwards. At 5 metres (16 ft) long, *Sauropelta* was an average size for a nodosaurid.

TRIASSIC 245-208 MYA	EARLY/MID JURASSIC 208-157 MYA	LATE JURASSIC 157-146 MYA	EARLY CRETACEOUS 146-97 MYA	LATE CRETACEOUS 97-65 MYA

EUOPLOCEPHALUS

The best known of the ankylosaurids is *Euoplocephalus*, 5 metres (16 ft) long and living a little earlier than *Ankylosaurus* in Alberta, Canada. Its back was an articulated mass of armour, the head an armoured box. Even the eyelids were armoured and slammed shut like the steel shutters of a battleship whenever danger approached. It used its powerful club tail to repel attack from even the most intimidating of predators, such as this *Tyrannosaurus rex*.

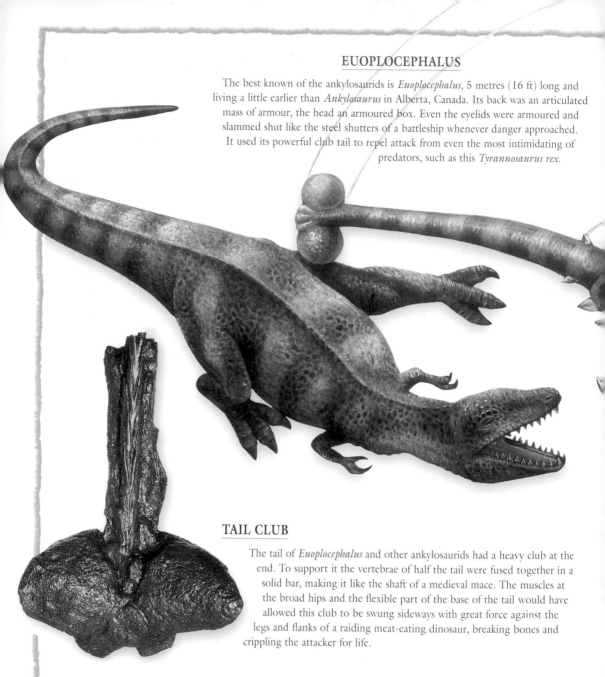

TAIL CLUB

The tail of *Euoplocephalus* and other ankylosaurids had a heavy club at the end. To support it the vertebrae of half the tail were fused together in a solid bar, making it like the shaft of a medieval mace. The muscles at the broad hips and the flexible part of the base of the tail would have allowed this club to be swung sideways with great force against the legs and flanks of a raiding meat-eating dinosaur, breaking bones and crippling the attacker for life.

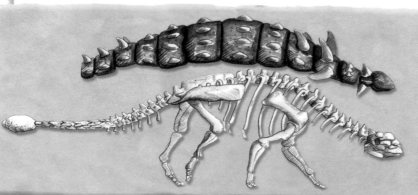

A HEARTY APPETITE

Without its armour we can see that *Euoplocephalus* was a heavy four-footed animal. Its hips were broad but the design allowed the guts to be carried well back. The guts would have been massive and probably contained fermenting chambers like those of modern cows.

THE ANKYLOSAURIDS
- THE CLUB-TAILS

The ankylosaurids were closely related to the nodosaurids but mostly came later, towards the end of the Cretaceous period. With their armoured necks and backs, they looked very like their relatives but instead of having spikes on the shoulders and the sides they had a different weapon – a heavy bony club at the end of the tail. This could have been devastating when swung at an enemy. It is also possible that the club may have been used as a decoy. Perhaps the club on the tail looked like a head on a neck, and caused meat-eaters to attack it instead of the more vulnerable front end.

CRETACEOUS UNDERGROWTH

By the end of the Cretaceous period, modern plants had evolved. Beneath the broad-leaved trees was an undergrowth of flowering herbs such as buttercups. The ankylosaurids and the nodosaurids carried their heads low and their mouths close to the ground. They were evidently low-level feeders that ate the flowering herbs.

TRIASSIC 245-208 MYA	EARLY/MID JURASSIC 208-157 MYA	LATE JURASSIC 157-146 MYA	EARLY CRETACEOUS 146-97 MYA	LATE CRETACEOUS 97-65 MYA

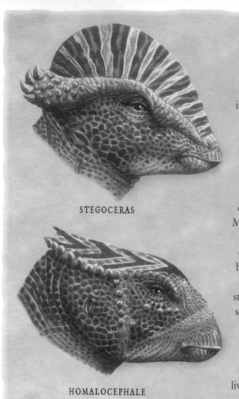

STEGOCERAS

HOMALOCEPHALE

HEADS

Each group of pachycephalosaurids had its own type of skull shape and ornamentation. *Stegoceras* and *Homalocephale*, from Mongolia, had sloping heads, higher at the rear, the latter with an elaborate head crest. *Prenocephale*, also from Mongolia, had a more rounded dome-like head. Both had decorative lumps around the bony crown. North-American *Stygimoloch* was perhaps the strangest, with a weird array of spikes and spines all around its dome. These were probably used for intimidation rather than fighting. All pachycephalosaurids lived in the late Cretaceous era.

STYGIMOLOCH

PRENOCEPHALE

MODERN SPARRERS

In the American Rocky Mountains, modern bighorn sheep and mountain goats go through an annual ritual in which the males fight the flock leader to test his strength. The construction of their skulls and horns protects them from suffering much damage when they bash against one another. Pachycephalosaurids probably had similar rituals.

TRIASSIC 245-208 MYA	EARLY/MID JURASSIC 208-157 MYA	LATE JURASSIC 157-146 MYA	EARLY CRETACEOUS 146-97 MYA	LATE CRETACEOUS 97-65 MYA

BONEHEADS

I magine a dinosaur, a two-footed plant-eating dinosaur such as an ornithopod, but give it a very high forehead so that it looks brainy. What you would have is a pachycephalosaurid – another dinosaur group descended from the ornithopods. The intelligent look is actually false. The brain in that head is tiny and the roof of the skull is made up of very thick bone. We think the mass of bone on top of the head was used as a weapon – a kind of battering ram. This was probably not for use against predators but for display purposes in courtship battles.

ALL CREATURES GREAT AND SMALL

There was a great range of sizes in pachycephalosaurids. The largest known, at 5 metres (16 ft) long, was North American *Pachycephalosaurus*. The smallest was *Micropachycephalosaurus* from China, which was about the size of a rabbit. This smallest of dinosaur has the longest dinosaur name ever given.

HORNED BATTLERS

The most complete pachycephalosaurid known is *Stegoceras*. The grain of its head bone and the strength of its neck probably provided an awesome degree of protection. They seem to have lived in herds. The adult males probably fought with one another to be the leader of the herd; the strongest of them would then mate with the females.

HARD HEADS

Dinosaur skulls are rarely preserved as fossils, but pachycephalosaurid skulls were different. The top bone of the skull was so massive it often survived as a fossil. Commonly, the only part of the animal to be preserved, these skulls are often found very battered. This suggests they were washed down a river for long distances before being buried in sediment. This may mean they were mountain-living animals, like their modern counterparts – mountain sheep.

THE PRIMITIVE-HORNED DINOSAURS

The last of the plant-eating dinosaur groups existed from the mid to late Cretaceous period. Like the ankylosaurids and the nodosaurids they lived in North America and in Asia, and they also evolved from ornithopods. They were equipped with armour but it was confined solely to the head. Early types were lightly-built and very ornithopod-like, but in later forms the armour on the head became so heavy they went about as four-footed animals. Flamboyant neck shields and horns evolved and these horned dinosaurs became known as the ceratopsians.

AN EARLY SHEEP

Scientists regard *Protoceratops* as the sheep of late Cretaceous Mongolia. Similar in size, they lived in herds and grazed the sparse vegetation of the arid landscape. One particular skeleton had the skeleton of a fierce carnivore, *Velociraptor*, clinging to its head shield. The meat-eater had attacked the ceratopsian with its killing claws, but the ceratopsian must have fought back with its big beak as both dinosaurs lost their lives during the fight.

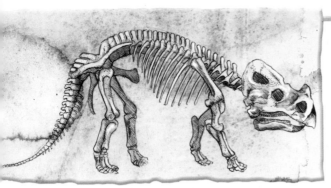

BIG BROTHER

By the time *Montanoceratops* had evolved, towards the end of the late Cretaceous, the ceratopsians were bigger and had developed the horns on their head. *Montanoceratops* was about 3 metres (10 ft) long and walked on all fours. However, like its two-footed ancestors, it still had claws on its feet. In later ceratopsians the toenails developed into hooves, better able to take the weight of big animals.

ARCHAEOCERATOPS

The most primitive of the ceratopsians known is *Archaeoceratops*. It was a very small animal, about a metre (3 ft) long, and scampered nimbly on hind legs about the plains of late Cretaceous China. It had a head that was very similar to that of *Psittacosaurus*. Its skeleton was so primitive and generalized it is possible that its descendants gave rise to the big ceratopsians that were to follow.

WHO'S A PRETTY BOY?

An early relative of the ceratopsians was the 1.5 metre (5 ft) long parrot-lizard *Psittacosaurus*. It developed a very strong beak and powerful jaws for plucking and chopping the tough vegetation it ate. A bony ridge around the back of the skull anchored its strong jaw muscles. The bony ridge and its big beak gave its skull a square shape, and the head must have looked very like the big-beaked head of a modern parrot.

CYCAD FOSSIL

At the end of the Cretaceous period the old-style vegetation was largely replaced by modern species of plants. However, some of the older types of palm-like cycads remained in some regions. In the areas where they occurred, the ceratopsians may have relied on these plants. Their narrow beaks would have reached into the palm-like clump of fronds and selected the best pieces, and their strong jaws would have shredded the tough leaves.

TRIASSIC 245-208 MYA	EARLY/MID JURASSIC 208-157 MYA	LATE JURASSIC 157-146 MYA	EARLY CRETACEOUS 146-97 MYA	LATE CRETACEOUS 97-65 MYA

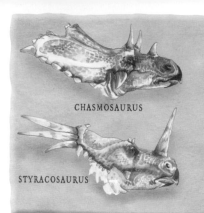

VARIETY OF HEADS

Ceratopsians all had the same body shape, but their different shapes of shield and horn arrangements made each type easily recognizable to their own herd. *Styracosaurus* had a monumental horn on its nose and an array of horns around its shield. *Chasmosaurus* had an enormous, sail-like shield. *Einiosaurus* had a long nose horn that curved forwards and a pair of straight horns at the edge of its shield. *Acheluosaurus* had a battering ram on its nose, a pair of short, blade-like horns above its eyes, and a curved pair at the shield-edge.

CHASMOSAURUS

STYRACOSAURUS

ALL FOR ONE

The horns of ceratopsians would have been used to defend themselves and their herd against big carnivores and also to tussle with one another position in the herd. Having locked horns, they would have pushed and shoved until one of them gave way. Little harm would have come to the loser. While travelling, the ceratopsians may have kept their young at the centre of the herd to protect them. If attacked by carnivores they may have formed a circle with the youngsters in the centre and the adults facing outwards so that the attackers were faced with the shields and horns of all the herd. Today, musk-oxen protect their herd in this way.

WILDEBEEST

We know ceratopsians moved about in herds because we have found bone beds consisting of many hundreds, even thousands, of skeletons. The animals would have been migrating, travelling in herds to areas where there was more food at a particular time of year. When crossing a river they may have been caught by a sudden flash flood that washed them away and dumped their bodies. This still happens in Africa today as herds of wildebeest migrate from one feeding ground to another.

THE BIG-HORNED DINOSAURS

The big ceratopsians were probably the most spectacular dinosaurs of the late Cretaceous period. They were all four-footed animals, mostly as big as today's rhinoceros. The ridge of bone around the neck had evolved into a broad shield. They also had an array of long horns on the face. The skulls of the big ceratopsians were so solid and tough that they were often preserved as fossils. As a result, we know a lot about their heads. There were two main lines of evolution. One group developed long frills and a pair of long horns above the eyes; the other had shorter frills and tended to have a single horn on the nose.

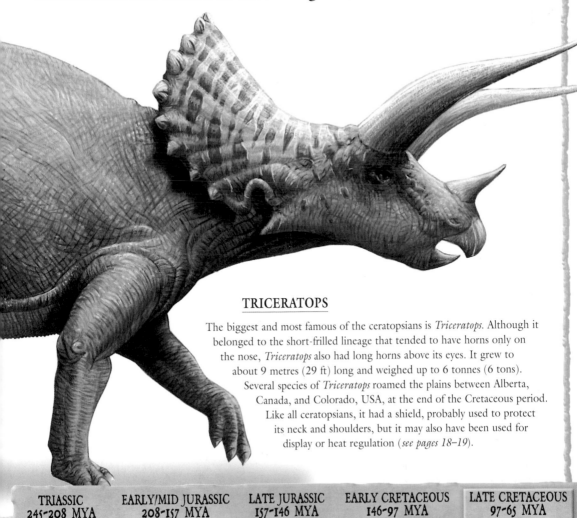

TRICERATOPS

The biggest and most famous of the ceratopsians is *Triceratops*. Although it belonged to the short-frilled lineage that tended to have horns only on the nose, *Triceratops* also had long horns above its eyes. It grew to about 9 metres (29 ft) long and weighed up to 6 tonnes (6 tons). Several species of *Triceratops* roamed the plains between Alberta, Canada, and Colorado, USA, at the end of the Cretaceous period. Like all ceratopsians, it had a shield, probably used to protect its neck and shoulders, but it may also have been used for display or heat regulation (*see pages 18–19*).

TRIASSIC 245-208 MYA	EARLY/MID JURASSIC 208-157 MYA	LATE JURASSIC 157-146 MYA	EARLY CRETACEOUS 146-97 MYA	LATE CRETACEOUS 97-65 MYA

DID YOU KNOW?

Dinosaurs are given long scientific names usually based on Latin or ancient Greek, so they can be recognized by scientists around the world no matter what language they themselves speak.

Scientific names for dinosaurs have two parts. Hence, *Tyrannosaurus rex*. The first part is the genus name and the second part is the species name. Usually, when not in a scientific context, only the genus name is used. However, the full name is sometimes so impressive that people use it all the time. If we do this we should also refer to *Triceratops* as *Triceratops prorsus*. After first referring to an animal's full scientific name, scientists subsequently shorten it by replacing the genus part with its initial. For example, *Tyrannosaurus rex* becomes *T. rex*. The scientific name is always printed in italics. The genus part always has a capital letter, while the species part does not. This holds true even if the name is based on the name of a person or a place. For example, *Protoceratops andrewsii*.

Dinosaur names sometimes change. Old books usually refer to *Apatosaurus* as *Brontosaurus*. Back in the nineteenth century,

specimens of it were found by two different people. The first was found in 1877 by collector Arthur Lakes who sent it to the palaeontologist Othneil Charles Marsh, who named it *Apatosaurus ajax*. Two years later, collector William Reed found an even better skeleton and sent it to Marsh who named it *Brontosaurus excelsus*. For almost a century they were thought to be two different animals, and *Brontosaurus* was the more famous. Eventually, it was realized they were the same. When a confusion like this occurs the rule is that the first name given is the only valid one, so the evocative *Brontosaurus* had to be dropped in favour of *Apatosaurus*.

Occasionally, a dinosaur is given a name that, unknown to the scientist who names it, has already been given to something else. When it was first studied in 1998, the bird/dinosaur *Rahonavis ostromi* was originally called *Rahona ostromi*. It had to be changed when it was noticed that the name Rohana had already been given to a beetle. *Mononychus* was changed to *Mononykus* for the same reason. A change of spelling is sometimes all that is required.

ACKNOWLEDGEMENTS

We would like to thank: Advocate, 'www.fossilfinds.com' Helen Wire and Elizabeth Wiggans
Lisa Alderson, Simon Mendez and Luis Rey.
Copyright © 2004 *ticktock* Entertainment Ltd.
First published in Great Britain by ticktock Publishing Ltd., Unit 2, Orchard Business Centre, No
All rights reserved.
No part of this publication may be reproduced, stored in a retrieval system, or transmitted in an
photocopying, recording or otherwise, without prior written permission
A CIP catalogue record for this book is available from the British Library. ISBN

Picture Credits:
t=top, b=bottom, c=centre, l=left, r=right, OFC=outside front cover, IFC=inside front cover, I

Lisa Alderson: 9br, 12b, 13b, 14b, 22-23c, 26-27c. John Alston: 2tl, 2tr, 2b, 7cr, 8t, 16b, 18t,
Z Botanical: 16bl, 17tr, 23cr. BBC Natural History Unit: 26b, 30b. Dr. Jose Bonaparte: 3t,
Museum: 8cl. National Trust: 7cl. Simon Mendez: 2-3c, 4t, 4-5c, 6-7c, 8-9b, 10b, 14-15c,
Museum of Utah: 24tl. Natural History Museum: 12t, 12cl, 13t, 14ct, 17t, 18b, 21cr, 22cl, 2
11c, 16-17c, 28-29c, 30cl. Professor Kent Stephens:

Every effort has been made to trace the copyright holders and we apologize in advanc
We would be pleased to insert the appropriate acknowledgement in any subsequ

snapping-turtle
guide